How to Set up YOUR "Perfect" Amateur Radio Go-kit

A guide for selecting, assembling, and utilizing a portable amateur radio station, away from your home station

By Scott Roberts, KK4ECR

February 2023

Table of Contents

Introduction

I was born in Washington D.C. and raised in California. From a young age, I had a fascination with Citizens Band and Amateur Radio. As I grew older, I decided to obtain my license to contribute to the community through public service. In 2011, I got my Technician Class license with the call sign of KK4ECR. One of the ways I chose to give back to the community was by assisting others in developing their experience and skills in the realm of Amateur Radio. During my interactions with other Amateur Radio operators, I became familiar with the concept of a "Go-Kit" and decided to construct one for myself.

For the past 11 years, I have been actively engaged in the construction of Go-Kits for myself and other Amateur Radio operators. My work has been recognized through interviews on various news stations and publications. Furthermore, I am a co-instructor for a class aimed at assisting individuals in obtaining their Amateur Radio Technician license (http://www.zero2licensed.com). Additionally, I have contributed to the industry through the publication of several articles in both Amateur Radio magazines and newsletters.

The objective of this book is to provide a comprehensive guide for assembling and utilizing a portable amateur radio station, referred to as a Go-Kit, in an efficient and cost-effective manner. While there is no "perfect" Amateur Radio Go-Kit, there can be "YOUR Perfect" Amateur Radio Go-Kit. The "perfect" go-kit is one that meets your needs – everyone has different needs and wants when it comes to assembling a go-kit.

The amount of time it will take to build your Go-Kit will vary (from a few hours to a couple of weeks), depending on the existing equipment and the additional equipment that may be required. Once the selection of the transceiver, antenna, and case for the kit has been made, the focus shifts to the assembly and organization of the gear. Have fun building your go-kit and make it a learning experience.

Before You Start

To operate in the Amateur Radio bands, it is essential to hold at least a Technician Class license. To assist with obtaining this license, many local Amateur Radio clubs offer support and resources. Although taking a class is not mandatory, it can provide valuable knowledge and preparation for the licensing exam. For those interested in online learning, we recommend the "Ham Radio Zero to License in 6 Hours" self-study course available at:

http://www.zero2licensed.com/self-study

This comprehensive program covers all material necessary for successfully passing the Technician Class license exam.

Building an Amateur Radio "Go-Kit" can be a lot of fun. Deciding what radio and antenna you want to use can be a great learning process for both new and experienced radio operators.

Every radio operator will start assembling your Go-Kit with different equipment and different experience levels. No matter where you are in the process, the one thing you will need is a willingness to learn and put in a little time each day (30 minutes to an hour) to complete the project. There is no "wrong way" to build a "Go-Kit." Remember that this is YOUR Go-Kit, and you can assemble it any way that fits your needs and situation.

Some tools that you may also want to have handy:

- Screw Drivers
- Soldering Iron (if you plan to make your own cables and connectors)
- Electrical Tape
- Zip Ties (to organize wires)

Other items you will want to have will vary depending on our desired Go-Kit design, and many of these are covered in the chapters of this book.

So, let's get started building your "Go-Kit."

Determine your budget and needs

The first step in setting up a ham radio go kit is to determine your budget and needs. Consider the following factors when making this determination:

Your Budget

The cost of a good amateur radio go kit can vary greatly depending on several factors, including the level of skill of the person building the kit and their specific needs. For someone just starting out in amateur radio, a basic kit can be put together for just a few dollars, using items that may already be on hand. This might include a simple handheld transceiver, a spare battery or two, and a basic antenna.

On the other hand, someone with more advanced skills and specific needs may opt for a more expensive kit that includes a higher-powered transceiver, a more sophisticated antenna, and a larger battery or power source. These kits can cost several hundred dollars or more. Additionally, you might want to add some more accessories to your kit such as a power supply, a portable antenna, a CW key, etc.

Ultimately, the cost of a good amateur radio go kit will depend on the individual's needs and preferences. A good starting point is to determine the specific requirements of the intended usage, and then research different options and compare prices to find the best fit for your budget.

Type of activities

There are a few different types of activities that you might use a ham radio go kit for, and the equipment you include in your kit will depend on the type of activities you plan to engage in.

Casual communication: If you plan to use your ham radio go kit for casual communication, you might include equipment such as a handheld transceiver, a microphone, and an antenna. You may also want to include some spare batteries or a charger, as well as any necessary cables or connectors.

Emergency situations: If you plan to use your ham radio go kit in emergency situations, you will want to include equipment that is more rugged and reliable. You might consider including a portable transceiver with a built-in battery, a hand-crank or solar-powered charger, and an emergency beacon. You will also want to make sure you have enough spare batteries to last for an extended period of time, as well as any necessary cables or adapters.

Both: If you plan to use your ham radio go kit for both casual communication and emergency situations, you will want to include a mix of equipment that is suitable for both types of activities.

Location

The location where you will be using your ham radio go kit is an important consideration when determining what equipment to include.

Urban areas: If you will be using your ham radio go kit in urban areas, you may not need as much range as you would in a remote location. In an urban environment, you are likely to be near other ham radio operators, so you may not need a high-  powered transceiver or a long-range antenna. Instead, you might choose to include a handheld transceiver with a shorter-range antenna, as well as any necessary cables or connectors.

Remote locations: If you will be using your ham radio go  kit in a remote location, you will likely need a transceiver with a longer range and a more powerful antenna. You may also want to include a spare battery or a charger, as well as any necessary cables or adapters.

Both: If you will be using your ham radio go kit in both urban and remote locations, you will want to include equipment that is suitable for both types of environments.

Frequency range

The frequency range of a transceiver is an important consideration when choosing equipment for your ham radio go kit. Ham radios operate on a range of frequencies, and different transceivers are designed to operate on specific frequency ranges.

When choosing a transceiver for your ham radio go kit, it is important to make sure that it operates on the frequencies that you need. If you plan to use your ham radio for local communication, you will want a transceiver that operates on the 2-meter and 70-centimeter bands, which are commonly used for local communication. If you plan to use your ham radio for longer-range communication, you will want a transceiver that operates on the 6-meter, 10-meter, or higher bands, which are better suited for long-range communication.

It is also worth noting that different countries have different regulations regarding the frequencies that can be used for ham radio communication. Make sure to familiarize yourself with the regulations in your country before choosing a transceiver.

By considering these factors, you can determine your budget and needs and make sure to include the necessary equipment in your ham radio go kit.

Choose a portable transceiver

A portable transceiver, or handheld radio, is the core component of a ham radio go kit. When choosing a transceiver, consider the following factors:

Frequency range

As mentioned above, make sure to choose a transceiver that operates on the frequencies you need.

Power output

The power output of a transceiver is an important factor to consider when choosing equipment for your ham radio go kit. The power output of a transceiver determines its range, with higher power output generally resulting in greater range. However, it is important to note that higher power output also means higher battery usage, so you will need to balance the need for range with the need for energy efficiency.

If you plan to use your ham radio for local communication, you may not need a transceiver with a high-power output. A handheld transceiver with a power output of 5 watts or less should be sufficient for most local communication needs. On the other hand, if you plan to use your ham radio for longer-range communication, you will likely need a transceiver with a higher power output.

A portable transceiver with a power output of 50 watts or more may be more suitable for long-range communication.

It is worth noting that different countries have different regulations regarding the maximum allowable power output for ham radio transceivers. Make sure to familiarize yourself with the regulations in your country before choosing a transceiver.

Durability

If you plan on using your ham radio go kit in rough or rugged conditions, it is important to choose a transceiver that is built to withstand those conditions. A transceiver with a tough exterior and good water resistance will be better able to withstand the wear and tear of outdoor use and will be less likely to fail in challenging environments.

There are a few key features to look for in a transceiver that is built to withstand rough or rugged conditions:

Tough exterior: A transceiver with a tough exterior, such as a ruggedized or mil-spec casing, will be better able to withstand impacts, drops, and other types of physical abuse.

Water resistance: If you plan to use your ham radio in wet or rainy conditions, it is important to choose a transceiver with good water resistance. Look for a transceiver with an

IPX rating of at least 4, which indicates that it is protected against splashing water from any direction.

Dust resistance: If you plan to use your ham radio in dusty or dirty environments, it is important to choose a transceiver with good dust resistance. Look for a transceiver with an IPX rating of at least 6, which indicates that it is protected against dust ingress.

Temperature resistance: If you plan to use your ham radio in extreme temperature environments, it is important to choose a transceiver that is resistant to extreme temperatures. Look for a transceiver with a wide operating temperature range, such as -20°C to +50°C (-4°F to 122°F).

By choosing a transceiver with these features, you can be confident that your ham radio will be able to withstand the rigors of outdoor use in rough or rugged conditions.

Select an antenna

An antenna is an important component of a ham radio go kit, as it determines the range and performance of your transceiver. There are several types of antennas to choose from, including the following:

Wire antennas

Wire antennas are a popular choice for ham radio go kits because they are simple and inexpensive, and they can be easily deployed in a variety of locations. However, it is worth noting that wire antennas are not as efficient as other types of antennas and may not provide as much range.

Wire antennas consist of a single length of wire that is suspended between two points, such as trees or poles. They are relatively easy to set up and can be used in a variety of locations, including in urban areas where space is limited. However, wire antennas are not as efficient as other types of antennas, such as Yagi or vertical antennas, and may not provide as much range.

Wire antennas are a good choice for ham radio go kits if you need a simple and inexpensive antenna that can be easily deployed in a variety of locations. However, if you need a more efficient antenna with a longer range, you may want to consider using a different type of antenna.

Vertical antennas

Vertical antennas are a type of ham radio antenna that is more efficient than wire antennas and can provide greater range. They consist of a vertical element, or "radiator," supported by a base and a counterpoise, or "ground plane." Vertical antennas are relatively easy to set up and can be used in a variety of locations, including in urban areas where space is limited.

One of the main advantages of vertical antennas is their efficiency. Because the radiating element is vertical, the antenna is more sensitive to signals coming from the horizon, which makes it more effective at long-range communication. Vertical antennas are also relatively easy to match to a transceiver, which makes them a good choice for use with a variety of different transceivers.

However, it is worth noting that vertical antennas are typically larger and more cumbersome than other types of antennas, such as wire antennas or Yagi antennas. They may also require a larger space to set up, and may be more difficult to transport in a ham radio go kit.

Overall, vertical antennas are a good choice for ham radio go kits if you need a more efficient antenna with a longer range, but you should be prepared for the added size and weight of this type of antenna.

Portable antennas

Portable antennas are a type of ham radio antenna that is designed to be easily carried and deployed in a variety of locations. They are typically smaller and more lightweight than other types

of antennas, which makes them a good choice for those who need a compact and portable antenna.

There are a few different types of portable antennas, including telescoping whip antennas, which are similar to the antenna on a handheld transceiver, and collapsible

"folding" antennas, which can be easily carried in a bag or backpack. Portable antennas can be a good choice for those who need an antenna that is easy to transport and can be set up quickly in a variety of locations.

However, it is worth noting that portable antennas may not provide as much range as other types of antennas, such as wire antennas or vertical antennas. They may also be less efficient and may not perform as well in certain types of environments.

Overall, portable antennas can be a good choice for ham radio go kits if you need a compact and lightweight antenna that is easy to transport and can be set up quickly in a variety of locations. However, you should be prepared for the potential trade-off in range and performance.

Add supporting equipment

Some of the equipment that you may want to add to your ham radio go kit:

Power supply

 A portable battery pack or solar panel can be a useful addition to a ham radio go kit, particularly if you will be using your ham radio in a location without access to electricity. These types of power supplies can provide a reliable source of power for your transceiver when you are on the go and can be particularly useful in emergency situations.

.When choosing a power supply for your ham radio go kit, it is important to consider the power requirements of your transceiver. You will need a power supply that can provide enough power to operate your transceiver for the duration of your activity. For example, if you plan to use your ham radio for an extended period, you will need a power supply with a higher capacity. On the other hand, if you only need to use your ham radio for a short period of time, a smaller capacity power supply may be sufficient.

It is also important to consider the size and weight of the power supply, as well as its charging time and durability. You will want to choose a power supply that is compact and lightweight, and that can be charged quickly and easily. A durable power supply that is resistant to impacts and water damage is also a good choice, particularly if you plan to use your ham radio in rough or rugged conditions.

Overall, a portable battery pack or solar panel can be a useful addition to a ham radio go kit, particularly if you will be using your ham radio in a location without access to electricity. Just make sure to consider the power requirements of your transceiver when choosing a power supply.

Cables and connectors

Cables and connectors are an important part of a ham radio go kit, as they allow you to connect your transceiver to other equipment, such as a computer or a power supply. It is a good idea to include a variety of cables and connectors in your go kit so that you can connect your transceiver to different types of equipment.

Some common types of cables and connectors to include in a ham radio go kit are:

USB cables: USB cables are used to connect a transceiver to a computer or other device with a USB port. They are commonly used to connect a transceiver to a computer for programming or to connect a transceiver to a power supply for charging.

Coaxial cables: Coaxial cables are used to connect a transceiver to an antenna or other RF equipment. They are typically used to connect a transceiver to a directional antenna, such as a Yagi antenna.

Power cables: Power cables are used to connect a transceiver to a power supply, such as a battery pack or a wall outlet. They are typically used to provide power to a transceiver when it is not being used with a battery.

It is also a good idea to include a variety of adapters and connectors in your go kit, such as coaxial connectors, USB adapters, and power connectors. This will allow you to connect your transceiver to different types of equipment and ensure that you have the necessary cables and connectors on hand when you need them.

Microphone

A microphone is an important piece of equipment for

making voice transmissions with a ham radio. Some transceivers come with a built-in microphone, while others require an external microphone. When choosing a microphone for your ham radio go kit, there are a few key factors to consider:

Sensitivity: The sensitivity of a microphone refers to its ability to pick up sound. A microphone with a high sensitivity will be able to pick up even quiet sounds, while a microphone with a low sensitivity may struggle to pick up quieter sounds. When choosing a microphone, consider the type of environment you will be using it in and the level of background noise you are likely to encounter.

Frequency response: The frequency response of a microphone refers to the range of frequencies that it can reproduce. A microphone with a wide frequency response will be able to reproduce a wide range of frequencies, from low bass to high treble. A microphone with a narrow frequency response may struggle to reproduce certain frequencies. When choosing a microphone, consider the type of voice transmissions you will be making and the range of frequencies you will need to reproduce.

Size and weight: If you are including a microphone in your ham radio go kit, you will want to choose a microphone that is compact and lightweight. This will make it easier to carry and transport and will allow you to easily access the microphone when you need to make a transmission.

Overall, a microphone can be a useful addition to a ham radio go kit, particularly if you plan to make voice transmissions. Just make sure to consider factors such as sensitivity, frequency response, and size and weight when choosing a microphone.

Headphones

Headphones can be a useful addition to a ham radio go kit, particularly if you need to listen to transmissions without disturbing others. Some transceivers come with a built-in speaker, while others require an external speaker or headphones. When choosing headphones for your ham

radio go kit, there are a few key factors to consider:

Comfort: When listening to transmissions for an extended period, comfort is an important factor to consider. Look for headphones with a comfortable and secure fit and consider the weight and size of the headphones.

Noise isolation: If you will be using your headphones in a noisy environment, noise isolation may be an important factor to consider. Look for headphones with good noise isolation, which will help to block out background noise and allow you to hear transmissions more clearly.

Frequency response: The frequency response of headphones refers to the range of frequencies that they can reproduce. Headphones with a wide frequency response will be able to reproduce a wide range of frequencies, from low bass to high treble. When choosing headphones, consider the type of transmissions you will be listening to and the range of frequencies you will need to reproduce.

Durability: If you will be using your headphones in rough or rugged conditions, durability may be an important factor to consider. Look for headphones with a tough exterior and good water resistance, which will be better able to withstand the wear and tear of outdoor use.

Overall, headphones can be a useful addition to a ham radio go kit, particularly if you need to listen to transmissions without disturbing others. Just make sure to consider factors such as comfort, noise isolation, frequency response, and durability when choosing headphones.

Field reference manual

A field reference manual is a useful resource to have in your ham radio go kit, as it can provide valuable information on how to operate your transceiver, as well as other useful information such as frequency charts and operating procedures.

Some field reference manuals are specific to a particular model of transceiver, while others provide more general information on ham radio operation. A field reference manual can be a helpful resource to have on hand when you are setting up and using your ham radio, as it can provide guidance on how to use your transceiver and troubleshoot any issues that may arise.

In addition to providing information on how to operate your transceiver, a field reference manual may also include information on other aspects of ham radio operation, such as:

Frequency charts: A frequency chart can provide  information on the frequencies that are allocated for ham radio use in your region, as well as the type of activity that is allowed on each frequency.

Operating procedures: A field reference manual may provide guidance on how to properly conduct yourself on the air, including how to make voice transmissions and how to use proper radio protocol.

Emergency procedures: A field reference manual may provide information on how to use your ham radio in an emergency, including how to establish communication and how to transmit messages.

Overall, a field reference manual can be a useful resource to have in your ham radio go kit, as it can provide valuable information on how to operate your transceiver and other aspects of ham radio operation. *Nifty Quick Reference* is one of the more popular Field Reference Guide publishers – you can search Amazon for Nifty Reference Guide to find the guide that fits your equipment and operating environment.

By adding these supporting items to your ham radio go kit, you can ensure that you have all the equipment you need to operate your transceiver effectively when you are on the go.

Add emergency equipment

If you are planning to use your ham radio go-kit in emergency situations, you will need to include some emergency equipment to help you stay safe and self-sufficient. Here are some specific items you will need to your ham radio go kit:

Flashlight

A flashlight can be a useful addition to a ham radio go kit, as it can be useful in a variety of emergency situations, such as power outages or natural disasters. When choosing a flashlight for your ham radio go kit, there are a few key factors to consider:

Brightness: A bright flashlight will be more effective at illuminating a dark environment and can be useful for signaling or signaling in an emergency. Look for a flashlight with a high lumen output, which indicates its brightness.

Durability: A durable flashlight is important, particularly if you will be using it in rough or rugged conditions. Look for a flashlight with a tough exterior and good water resistance, which will be better able to withstand the wear and tear of outdoor use.

Battery life: A flashlight with a long battery life is  important, as you may need to use it for an extended period of time in an emergency situation. Look for a flashlight with a high-capacity battery or one that can use readily available batteries, such as AA or AAA batteries.

Size and weight: If you are including a flashlight in your ham radio go kit, you will want to choose a flashlight that is compact and lightweight. This will make it easier to carry and transport and will allow you to easily access the flashlight when you need it.

Overall, a flashlight can be a useful addition to a ham radio go kit, particularly if you will be using your ham radio in emergency situations. Just make sure to choose a flashlight that is bright, durable, and has a long battery life.

 **First aid kit**

A first aid kit can be an essential item to include in a ham radio go kit, particularly if you will be using your ham radio in emergency situations. A first aid kit can provide necessary supplies and equipment to treat injuries or illnesses in the field and can be a valuable resource in a variety of emergency situations.

When choosing items to include in a first aid kit for your ham radio go kit, it is a good idea to include a variety of items such as:

Bandages: Bandages can be used to cover and protect cuts and wounds. Consider including a variety of sizes and types of bandages, such as adhesive bandages, gauze pads, and roller bandages.

Gauze: Gauze can be used to wrap wounds and protect injuries. Consider including a variety of sizes and types of gauze, such as gauze pads and gauze rolls.

Adhesive tape: Adhesive tape can be used to hold bandages and gauze in place and provide additional support to injuries.

Pain relievers: Pain relievers, such as acetaminophen or ibuprofen, can be useful for relieving pain and reducing inflammation.

Antiseptic: Antiseptic products, such as alcohol wipes or hydrogen peroxide, can be used to clean and disinfect wounds.

Splints: Splints can be used to immobilize and support injuries, such as fractures.

Scissors and tweezers: Scissors and tweezers can be useful for cutting bandages and removing splinters or other small objects from wounds.

Overall, a first aid kit is an essential item to have in a ham radio go kit, as it can be used to treat minor injuries or provide basic medical care in an emergency situation. It is important to make sure that your first aid kit is well-stocked and includes a variety of items. By including a first aid kit in your ham radio go kit, you can be better prepared to deal with any medical emergencies that may arise while you are on the go.

Emergency food and water supplies

In an emergency situation, it is important to have a supply of food and water to sustain you. This is especially true if you will be using your ham radio go kit in a remote location or in a situation where you may be cut off from access to food and water.

To ensure that you have a sufficient supply of food and water in an emergency situation, it is a good idea to include non-perishable food items in your ham radio go kit. Some options for non-perishable food items include:

Energy bars: Energy bars are a convenient and portable source of nutrition that can help to sustain you in an emergency situation. They typically have a long shelf life and are easy to store.

Dehydrated meals: Dehydrated meals are a convenient way to store food that has a longer shelf life and takes up less space than other types of food. They are easy to prepare and can be a good source of nutrition in an emergency situation.

Jerky: Jerky is a type of preserved meat that has a long shelf life and is easy to store. It can be a good source of protein in an emergency.

In addition to non-perishable food items, it is also important to have a supply of clean water in an emergency situation. There are a few different options for obtaining clean water in an emergency:

 Water filtration system: A water filtration system, such as a water bottle with a built-in filter, can be used to filter water from sources such as rivers or lakes. This can be a good option if you are in an area where there is a sufficient supply of water, but it may not be safe to drink.

Water purification tablets: Water purification tablets can be used to purify water from sources such as rivers or lakes. These tablets contain chemicals that kill harmful bacteria and viruses, making the water safe to drink.

Overall, it is important to have a supply of food and water in an emergency, and it is a good idea to include non-perishable food items and a water filtration system or water purification tablets in your ham radio go kit. This will help to ensure that you have a sufficient supply of food and water to sustain you in an emergency.

By including these emergency items in your ham radio go kit, you can be better prepared to handle a variety of emergency situations. It is a good idea to regularly check and replace any expired or used items in your emergency kit to ensure it is always ready for use.

Organize and pack your go kit

 Once you have all the equipment you
need, it is important to organize and pack
your go kit properly. This will make it
easy to access the items you need when you are on the go.
Here are some specific tips for organizing and packing your
go kit:

Use a portable bag or case

A portable bag or case can be a useful addition to a ham
radio go kit, as it can help protect your equipment and
keep it organized. When choosing a bag or case for your
ham radio go kit, there are a few key factors to consider:

Durability: A durable bag or case is important, particularly
if you will be using your ham radio go kit in rough or
rugged conditions. Look for a bag or case with a tough
exterior and good water resistance, which will be better
able to withstand the wear and tear of outdoor use.

Size and capacity: Make sure to choose a bag or case that
is large enough to hold all your equipment. Consider the
size and shape of your equipment, as well as any
additional items you may need to carry, such as a
flashlight or first aid kit.

Organization: Look for a bag or case that has a good
organizational system, such as pockets, dividers, or

compartments, which can help to keep your equipment organized and easily accessible.

Portability: If you will be using your ham radio go kit on the go, it is important to choose a bag or case that is portable and easy to carry. Look for a bag or case with comfortable and adjustable shoulder straps or a handle, which will make it easier to transport.

A portable bag or case can be a useful addition to a ham radio go kit, as it can help protect your equipment and keep it organized. Just make sure to choose a bag or case that is durable, has enough space to hold all your equipment, and is portable.

Label everything clearly

Labeling your equipment clearly can be a useful way to keep your ham radio go kit organized and make it easier to find the items you need quickly and easily. There are a few different options for labeling your equipment, including:

Labels: Pre-made labels can be a convenient and easy way to label your equipment. Look for labels that are durable and water-resistant, which will be able to withstand the wear and tear of outdoor use.

Markers: A permanent marker or pen can be used to label your equipment directly. Make sure to use a marker that is waterproof and resistant to fading, which will ensure that the labels remain legible even if the equipment gets wet or is exposed to the elements.

Stickers: Stickers can be a fun and easy way to label your equipment. Look for stickers that are durable and water-resistant, which will be able to withstand the wear and tear of outdoor use.

No matter what method you choose, it is important to clearly and legibly label your equipment so that you can easily identify each item. Consider labeling items such as your transceiver, antennas, cables, and other equipment so that you can easily find what you need in an emergency.

Labeling your equipment clearly can be a useful way to keep your ham radio go kit organized and make it easier to find the items you need quickly and easily.

Arrange your equipment in a logical order

Arranging your equipment in an orderly fashion can be a useful way to keep your ham radio go kit organized and make it easier to find the items you need quickly and easily. There are a few different approaches you can take when organizing your equipment:

By function: You might want to arrange your equipment by the function it serves. For example, you might want to pack your transceiver and antenna first, followed by your cables and connectors, and then your supporting equipment such as a flashlight or first aid kit.

By frequency of use: You might want to arrange your equipment in an order that reflects how frequently you use each item. For example, you might want to pack the items you use most frequently, such as your transceiver and microphone, near the top of your bag or case for easy access.

By size: You might want to arrange your equipment by size, with the larger items at the bottom of your bag or case and the smaller items on top. This can help to prevent the smaller items from getting lost or buried beneath the larger ones.

The most important thing is to arrange your equipment in a way that makes sense to you and that allows you to easily access the items you need. By organizing your equipment in a logical and orderly fashion, you can save time and reduce frustration when you are using your ham radio go kit on the go.

Pack items that you may need to access quickly in an easily accessible location

If you anticipate needing to access certain items quickly in an emergency, it is a good idea to pack them in an easily accessible location in your ham radio go kit. This can help to ensure that you can quickly and easily access the items you need when you need them.

Some items that you might want to pack in an easily accessible location in your ham radio go kit include:

Flashlight: A flashlight can be an essential tool in an emergency, particularly if you need to navigate in low light conditions. Pack your flashlight in a location where you can easily grab it, such as a side pocket or an outer compartment of your bag or case.

First aid kit: A first aid kit is an essential item to have in an emergency, as it can be used to treat minor injuries or provide basic medical care. Pack your first aid kit in a location where you can easily grab it, such as a side pocket or an outer compartment of your bag or case.

Transceiver: Your transceiver is the main piece of equipment in your ham radio go kit, and you will likely need to access it frequently. Pack your transceiver in a location where you can easily grab it, such as a top pocket or an outer compartment of your bag or case.

If you anticipate needing to access certain items quickly in an emergency, it is a good idea to pack them in an easily accessible location in your ham radio go kit. This will help to ensure that you can quickly and easily access the items you need when you need them.

If you these tips, you can organize and pack your ham radio go kit in a way that makes it easy to access the items you need when you are on the go.

Test and practice

Before you head out with your ham radio go kit, testing and practicing with your ham radio go kit is an important step in ensuring that you are prepared to use it efficiently and effectively in any situation. Here are some specific tips for testing and practicing with your go kit:

Test your equipment

Before you head out with your ham radio go kit, it is important to test all your equipment to make sure it is working properly. This will help to ensure that you are able to use your ham radio effectively in a variety of situations.

Here are some steps you can follow when testing your ham radio go kit:

 Test your transceiver: Make sure that your transceiver is powered on and functioning properly. Check that all buttons and controls are working correctly, and that the display is functioning properly.

Test your antenna: Make sure that your antenna is properly connected to your transceiver, and that it is functioning correctly. Check that the antenna is properly aligned and that it is receiving and transmitting signals as expected.

Test your power supply: Make sure that your power supply is working properly and that it is providing enough power to your transceiver. If you are using a battery pack or solar panel, make sure that it is charged and ready to use.

Test any other equipment: Make sure to test any other equipment that you have included in your ham radio go kit, such as cables and connectors, microphone, and headphones. Make sure that each item is functioning properly and that it is compatible with your transceiver.

Test all equipment in combination: Make sure to test all your equipment in combination, to ensure that everything is working together properly. This might include transmitting and receiving signals using your transceiver and antenna or using your cables and connectors to  connect your transceiver to other equipment.

Following these steps, can help to ensure that your ham radio go kit is in good working order and that you are prepared to use it effectively in a variety of situations.

Practice using the equipment

In addition to testing your equipment, it is important to practice using it to become familiar with its functions and controls. This will help you operate your ham radio go kit efficiently and effectively when you are on the go and will enable you to use your ham radio more confidently in a variety of situations.

Here are some tips you can use when practicing with your ham radio go kit:

Familiarize yourself with the functions and controls of your transceiver: Take some time to become familiar with the functions and controls of your transceiver, including buttons, switches, and the display. Practice using your transceiver to transmit and receive signals, and to change settings such as the frequency or power output.

Practice using your antenna: Practice using your antenna  to transmit and receive signals, and to adjust its orientation as needed. Make sure to become familiar with the different types of antennas available and their specific characteristics, as this will help you choose the right antenna for the situation.

Practice using other equipment: Practice using any other equipment you have included in your ham radio go kit, such as a microphone, headphones, or a field reference manual. Make sure to become familiar with the functions and controls of each item, and how to use them in combination with your transceiver.

Practice with a fellow ham operator: Consider practicing with a fellow ham operator to get a feel for using the equipment in a real-world scenario. This can help you become more comfortable and confident with your ham radio go kit and will enable you to troubleshoot any issues that may arise.

Practicing with your ham radio go kit is an important step in becoming familiar with its functions and controls, and in using it efficiently and effectively when you are on the go.

Familiarize yourself with operating procedures

Knowing proper operating procedures is important for efficient and effective use of your ham radio go kit. Proper operating procedures can help you communicate more effectively and can help to ensure the safety and reliability of your communications.

Here are some steps you can follow to familiarize yourself with proper operating procedures:

Learn the basic procedures for making voice and data transmissions: Make sure to familiarize yourself with the basic procedures for making voice and data transmissions using your ham radio. This might include procedures for setting up your transceiver, establishing contact with other operators, and exchanging information.

Learn any specific procedures for your equipment: Some types of equipment, such as antennas or power supplies, may have specific procedures that you need to follow when using them. Make sure to familiarize yourself with these procedures to ensure that you are using your equipment safely and effectively.

Familiarize yourself with emergency procedures: Make sure to familiarize yourself with the proper procedures for handling emergency situations, such as natural disasters or accidents. This might include procedures for transmitting emergency messages or signals, or for seeking help from other operators.

Learn about good operating practices: Make sure to familiarize yourself with good operating practices, such as using proper language and etiquette when communicating with other operators. This can help to ensure that your communications are professional and respectful and can help to maintain the integrity of the ham radio community.

These steps will help you become familiar with the proper operating procedures for your ham radio go kit, which will enable you to use it efficiently and effectively when you are on the go.

By testing and practicing with your ham radio go kit, you can ensure that you are prepared to use it effectively in any situation. This will give you the confidence and skills you need to communicate effectively when you are on the go.

Author's Recommendations

Over the past seven chapters, we've outlined the steps to help you create YOUR "Perfect" Go-Kit. However, we want to provide you with additional value by sharing the author's top recommendations for three distinct Go-Kits, complete with links to where you can purchase each item. You'll also find a variety of Go-Kit designs featured at the end of this section.

Let's break this down into these Go-kits:

- *Go-Kit #1 – The Basic*
- *Go-Kit #2 – The Advanced*
- *Go-Kit #3 – The Extreme*

Followed by:

- *What should be included in all Go-Kits?*
- *What additional options are there for all Go-Kits?*

Go-Kit #1 – The Basic

- Case: Ammo-can or Small Toolbox
    - Ammo-can – Harbor Freight
      https://www.harborfreight.com/metal-050-caliber-ammo-can-63750.html
    - Small Toolbox – Harbor Freight
      https://www.harborfreight.com/tactical-ammoutility-box-64113.html

- Dual Band Mobile Radio
  - Yeasu FTM-6000R (or comparable radio)
    - Gigaparts
      https://www.gigaparts.com/yaesu-ftm-6000r-dual-band-mobile-transceiver.html
- Battery & Charger
  - 15 Amp-hour – BLF-1215A - Gigaparts
    https://www.gigaparts.com/bioenno-power-lifepo4-battery-blf-1215a.html
  - Bioenno Power 14.6V 6A AC-to-DC Charger BPC-1506A - Gigaparts
    https://www.gigaparts.com/bioenno-power-14-6v-6a-ac-to-dc-charger-bpc-1506a.html
- Antenna
  - Roll-up J-Pole
    - Build it yourself - ARRL Article
      https:/www.arrl.org/files/file/Public Service/TrainingModules/jpole-dual-band.pdf
    - Already built - Amazon.com
      https://a.co/d/7D7SgrY

Go-Kit #2 – The Advanced

- Case
  - Pelican 1600 - Amazon
    https://a.co/d/aubouiv
  - Nanuk 930 - Gigaparts
    https://www.gigaparts.com/nanuk-930-case-w-foam-black.html
  - Apache 4800 - Harbor Freight
    https://www.harborfreight.com/material-

handling/parts-storage/utility-cases-ammo-boxes/protective-cases/4800-weatherproof-protective-case-x-large-tan-56864.html

- All-Band / All-Mode Radio
  - o Yeasu FT-991a - Gigaparts
    https://www.gigaparts.com/yaesu-ft-991a.html
- Battery & Charger
  - o 30Ah LifePo4 ZBP-BLF-1230A – Gigaparts
    https://www.gigaparts.com/bioenno-power-blf-1230a-30ah-lifepo4-battery.html
  - o Bioenno Power 14.6V 10A AC-to-DC Charger BPC-1510A - Gigaparts
    https://www.gigaparts.com/bioenno-power-14-6v-10a-ac-to-dc-charger-bpc-1510a.html
- Antenna
  - o Roll-up J-Pole
    - Build it yourself - ARRL Article
      https:/www.arrl.org/files/file/Public Service/TrainingModules/jpole-dual-band.pdf
    - Already built - Amazon.com
      https://a.co/d/7D7SgrY
  - o G5RV Dipole Antenna (HF)
    - MFJ-1778M - Gigaparts
      https://www.gigaparts.com/mfj-1778m.html

Go-Kit #3 – The Extreme

- Case
  - o Gator 6U Case on Wheels – Amazon
    https://a.co/d/7poYc1p Also, you will want at

least one shelf on which to mount equipment –
Amazon https://a.co/d/0OVRPHL
- All-Band / All-Mode Radio
  - o Yeasu FT-991a - Gigaparts
    https://www.gigaparts.com/yaesu-ft-991a.html
- Battery & Charger
  - o 50Ah LifePo4 ZBP-BLF-1250A – Gigaparts
    https://www.gigaparts.com/bioenno-power-blf-1250a-50ah-lifepo4-battery.html
  - o Bioenno Power 14.6V 10A AC-to-DC Charger
    BPC-1510A - Gigaparts
    https://www.gigaparts.com/bioenno-power-14-6v-10a-ac-to-dc-charger-bpc-1510a.html
- Antenna
  - o Roll-up J-Pole
    - Build it yourself - ARRL Article
      https:/www.arrl.org/files/file/Public
      Service/TrainingModules/jpole-dual-band.pdf
    - Already built - Amazon.com
      https://a.co/d/7D7SgrY
  - o Any of these for HF:
    - Chameleon CHA Emcomm III –
      Gigaparts
      https://www.gigaparts.com/cha-emcomm-iii-base-antenna-10-80m.html
    - BuddiPole Mini - Buddipole
      https://www.buddipole.com/minibuddipole.html
    - BuddiStick PRO Deluxe - Buddipole
      https://www.buddipole.com/budepa.html

These should be included in all Go-Kits

In addition to the items listed with each Go-Kit option, these items should be included with every Go-Kit:

- Power Supply
  - Samlex SEC-1235M - Gigaparts
    https://www.gigaparts.com/samlex-sec-1235m.html
  - Alinco DM-330MVT - Gigaparts
    https://www.gigaparts.com/alinco-dm-330mvt.html
- Coax
  - 75 Foot – 218XATC-PL-75 - Gigaparts
    https://www.gigaparts.com/75ft-rg8x-lmr-240uf-coax-jumper-w-pl259-ends.html
  - *You will want one length of coax per antenna that is connected to the radio. (i.e. The Yeasu FT-991a allows for one HF antenna and one UHF/VHF) to be connected at one time.)*
- Paracord (100 foot)
  - Paracord can be obtained from many sources. Check out Amazon for the multiple options that they have.

o Paracord is extremely useful when putting your

Additional Tip: You will want to take some time to learn how to tie proper knots to securely hold your antennas. Check out this video on a great knot to use when setting up antennas – YouTube https://www.youtube.com/watch?v=gvzpxL4Y_4o

---

wire antennas up in trees or on antenna masts.

These are options for all Go-Kits

- Solar Panels – These are useful to keep your batteries charged and allow for longer operating in the field.
  - o Bioenno 120W (ZBP-BSP-120) – Gigaparts https://www.gigaparts.com/bioenno-power-bsp-120-foldable-solar-panel.html
  - o Thunderbolt Solar 100W – Harbor Freight https://www.harborfreight.com/100-watt-amorphous-solar-panel-kit-63585.html
- Solar Controller – This is needed for any solar panel to control the charging of your battery. *NEVER connect your solar panels directly to your battery.*
  - o Genasun 10a LifePo4 Controller - Gigaparts https://www.gigaparts.com/genasun-10a-mppt-solar-charge-controller-for-lifepo4-batteries.html
- Antenna Masts – You can use masts to setup your antennas if there are no trees to use for antennas.

- o Fiberglass Military Poles (used) – Google Search for multiple sources. You can also get these at most military surplus stores.
- Multi-Tool – A good multi-tool will come in very handy when setting up antennas and can be a great asset when needing to make minor repairs to your gear.
  - o Leatherman Bond Multitool – Amazon https://a.co/d/3ZPv2yB

Photos

Go-Kit #1 – **The Basic**

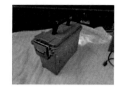  

Go-Kit #2 – **The Advanced**

Go-Kit #3 – The **Extreme**

Various setups in the field

Frequently Asked Questions

Here are a few questions that people most commonly ask about setting up their own Amateur Radio Go-Kit.

"Where do I get all that stuff for my go-kit?"

Here is a list of resources where you can go to acquire most of the things you will need for your go-kit.

- Study for your Ham Radio technician license: http://www.zero2licensed.com/self-study
- Take and pass your technician license test: https://www.laurelvec.com/
- Ham Radio transceivers, antennas, power supplies, etc:
  - Gigaparts - http://www.gigaparts.com
  - Ham Radio Outlet - https://www.hamradio.com/
  - DX Engineering - https://www.dxengineering.com/
- Microphones & Headsets
  - Heil Headsets - https://heilhamradio.com/products/headsets/
  - Heil Mics - https://heilhamradio.com/products/microphones
- Other items for your kit such as flashlights, first aid kits, etc.:
  - Amazon – http://www.amazon.com

"Where can I get help when I have questions?"

For assistance and access to a plethora of knowledge from experiences amateur radio operators, find your local amateur radio clubs. Utilize this helpful resource to advance your interests and talents. To locate a club in your area, you can go to http://www.arrl.org/find-a-club.

You can also email the author at kk4ecr@gmail.com.

"I built my Go-Kit, now what?

The best thing to do now, is to join a local club and gather as much knowledge as you can. You can also check http://www.zero2licensed.com/university for a list of online resources and classes to help you learn how to use your radio equipment.

What is Next?

Congratulations on taking the steps to build your Amateur Radio Go-Kit. Portable operations can be very rewarding. Find various ongoing outdoor amateur radio activities and put your efforts to good use. Here are just a few of the many portable amateur radio operating events:

- Winter Field Day (https://www.winterfieldday.com/)
- Parks on the Air (https://parksontheair.com/)
- Field Day (http://www.arrl.org/field-day)

Also, go to http://kk4ecr.us/wp/go-kit-bonus for three free bonus documents:

- BONUS 1 - Coax used for Amateur Radio
- BONUS 2 - Antennas Connectors
- BONUS 3 - Amateur Radio Power Connectors

You can also scan the QR Code below to get these files:

Conclusion

To summarize this book, setting up a good ham radio go kit involves several steps:

Determine your budget and needs: It is important to determine your budget and the type of activities for which you will be using the kit to determine the equipment you need to include.

Choose a portable transceiver and antenna: The transceiver is the main piece of equipment in your ham radio go kit, and it is important to choose a transceiver that meets your needs in terms of frequency range, power output, and durability. The antenna is also an important factor in the performance of your ham radio go kit, and it is important to choose an antenna that is suitable for the type of activities you will be using the kit for.

Add supporting equipment: There are a variety of other pieces of equipment that you may want to include in your ham radio go kit, such as cables and connectors, a microphone, headphones, and a field reference manual.

Consider emergency equipment: In an emergency situation, it is important to have a supply of food and water, as well as other emergency equipment such as a flashlight and first aid kit.

Organize and pack your go kit: It is important to arrange your equipment in an orderly fashion and pack it in a bag or case that is durable and has enough space to hold all your equipment.

Test and practice with your equipment: It is important to test and practice with your equipment to become familiar with its capabilities and limitations, and to ensure that it is in good working order.

With a little work you can assemble a reliable and effective ham radio go kit that will serve you well in a variety of situations.

When building your Go-Kit, keep this in mind:

There is no PERFECT Go-Kit, but there is "YOUR 'Perfect' Go-Kit."

Made in United States
North Haven, CT
27 September 2024

58021423R00031